S0-DYE-281

Life in the World's Biomes

Wetland Plants

by Terri Sievert

Consultant:
Ian A. Ramjohn, PhD
Department of Botany and Microbiology
University of Oklahoma
Norman, Oklahoma

Mankato, Minnesota

Bridgestone Books are published by Capstone Press,
151 Good Counsel Drive, P.O. Box 669, Mankato, Minnesota 56002.
www.capstonepress.com

Printed in the United States of America

Library of Congress Cataloging-in-Publication Data
Sievert, Terri.
Wetland plants / by Terri Sievert.
p. cm.—(Bridgestone Books. Life in the world's biomes)
Summary: "Describes wetland plants and wetland plant uses"—Provided by publisher.
Includes bibliographical references and index.
ISBN 0-7368-4325-6 (hardcover)
1.Wetland plants—Juvenile literature. I. Title. II. Series: Life in the world's biomes.
QK938.M3S45 2006
581.7'68—dc22 2004028516

Editorial Credits
Amber Bannerman, editor; Jennifer Bergstrom, designer; Kelly Garvin, photo researcher; Scott Thoms, photo editor

Photo Credits
Ann & Rob Simpson, 8
Bruce Coleman Inc./C.C. Lockwood, 4; Ed R. Degginger, 20; Janis Burger, 18; Kike Calvo V&W, 10; Lee Rentz, 12; Masa Ushioda, 6 (left); W.H. Black, 6 (top right)
Corbis/Phil Schermeister, 14; Sally A. Morgan, Ecoscene, 16
Digital Vision, 1
Image Ideas, 6 (bottom right)
Tom Till, cover

1 2 3 4 5 6 10 09 08 07 06 05

Table of Contents

Wetlands

Brown cattails poke out of a watery **marsh**. They look like hot dogs on a stick. Water lilies float lazily on the water's surface. Wetlands are full of plant life.

Wetlands can be found all over the world. Some wetlands have squishy soil people can walk on. Other wetlands are completely covered with water.

The water in wetlands comes from freshwater or saltwater sources. Freshwater wetlands are near streams or lakes. Saltwater wetlands are near seashores.

◂ Brown cattail and water lily roots stick to the bottom of marshes.

Wetland Plants

Freshwater marshes, **bogs**, and **swamps** are full of plants. Cattails and water lilies are plants that grow in freshwater marshes. Peat moss covers freshwater bogs. Cypress trees tower over freshwater swamps.

Fewer types of plants can **survive** in saltwater marshes and swamps. Some plants can't grow in the salty water and soil found in these areas. Other plants grow very well there. Tough cordgrass grows in saltwater marshes. Mangrove trees also live in saltwater swamps.

Mangrove trees (left), peat moss (top right), and water lilies (bottom right) all grow in wetlands.

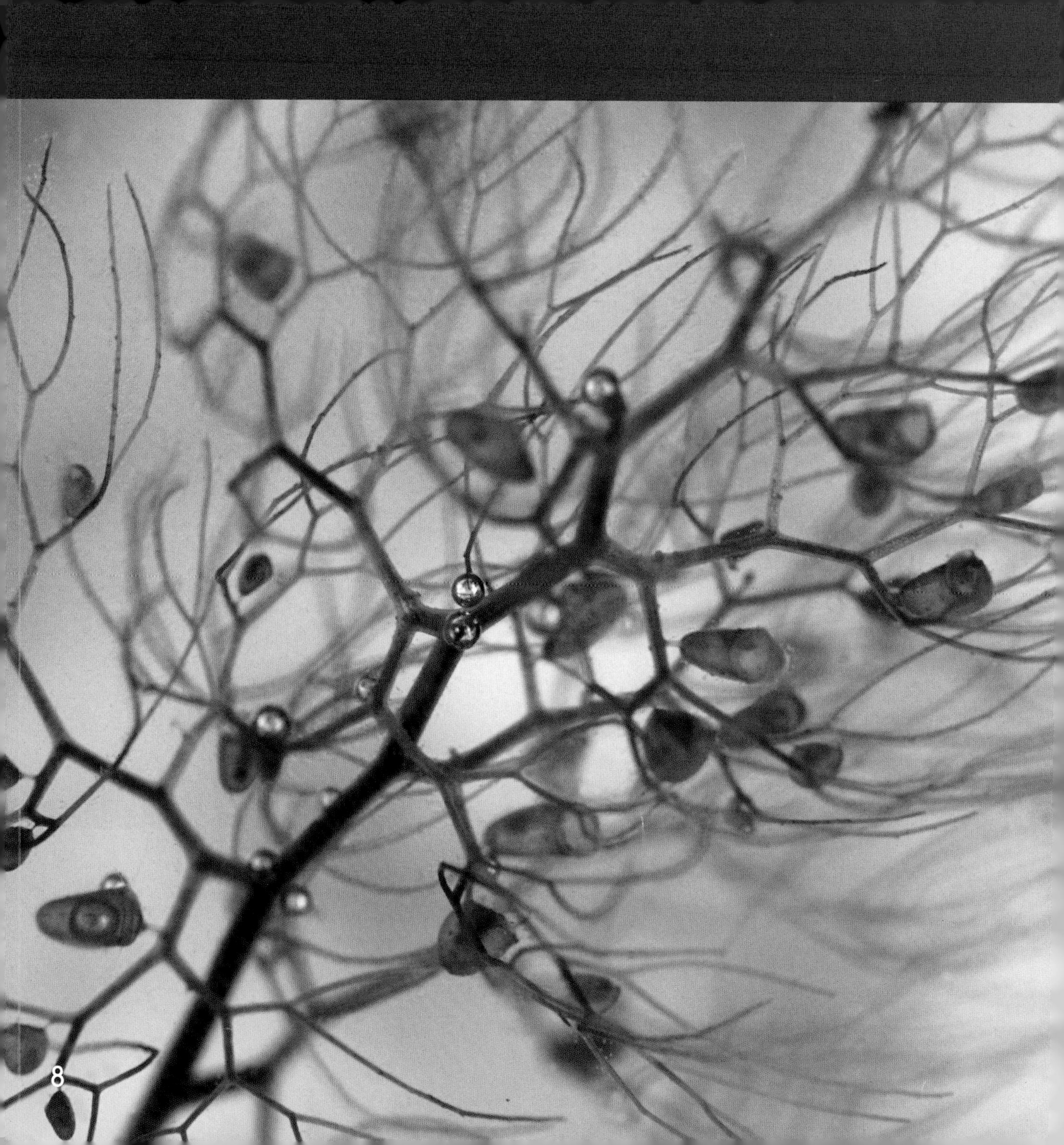

Wetland Plant Features

Underwater wetland plants often have thin leaves that move easily in the water. Water milfoil and bladderwort have strong stems that don't break in moving water.

Plants need **oxygen** to live. Some water in wetlands has little oxygen. Plants like the cypress tree grow roots that stick out of the water. The roots get oxygen from the air. Scouring rush plants live in damp soil near streams and lakes. These plants have hollow stems that take in oxygen. Oxygen travels through the stems to the roots.

◀ The bladderwort's leaves sway in the water. Its tiny pouches trap insects and baby fish.

Plant Homes for Animals

A variety of animals make their homes on mangrove trees. Tree crabs live on mangrove trunks. Young snapper fish hide underwater among mangrove roots. Sponges and sea squirts stick to hard mangrove roots.

Mammals and birds make homes from cattail plants. Muskrats build their lodges with cattail leaves. After the muskrats leave, minks often move into the lodges. Ducks and geese build nests out of cattail stems.

◆ A tree crab's sharp, pointed legs help it climb and live on mangrove trees.

Plant Foods for Animals

Furry animals eat wetland plants. Beavers eat the roots of water lilies. Muskrats feed on cattails and the softstem bulrush.

Birds also find plenty of food in wetlands. Stems, seeds, and leaves of water plants make good meals for Canada geese. Mallards eat wild rice. Swans pull and eat water milfoil and eelgrass plants from the wetland bottom. Floating plants like duckweed make good snacks for birds.

◀ Beavers use their claws to dig up water lily roots. Their strong front teeth help them chew the roots.

Plants Used by People

People can eat many wetland plants. Some people gather and eat wild rice that grows in freshwater marshes. Cranberries from bogs are part of most Thanksgiving meals. Arrowhead roots can also be eaten. These roots look like potatoes.

Peat moss is a useful bog plant. Gardeners use it as **mulch** to help other plants grow. People burn it as fuel. Peat moss can also be used to soak up oil spills in water. During World War I (1914–1918), peat moss was placed on soldiers' wounds to stop bleeding.

People use sticks to knock wild rice grains into a canoe.

Plants in Danger

When wetlands around the world are destroyed, so are the plants that live there. About half of the wetlands in the United States have been destroyed. Farms, homes, and businesses have been built on drained wetlands. In Wisconsin, drainage has caused the false asphodel plant to become a **threatened** plant.

Some wetland marshes have been deepened so they can be used for boating and waterskiing. Plants used to the shallow water die in the deep water.

◀ A mangrove swamp in Malaysia is cleared to make way for new buildings.

Protecting Wetland Plants

Governments pass laws to protect wetlands. These laws control how land in and near wetlands can be used. The Florida Everglades is protected as a national park. The wetlands there are being **restored**.

The U.S. government buys wetlands to restore the plants and animals. Ducks, geese, and other birds can build nests there. When hunters buy Federal Duck Stamps, the money is used to protect wetlands. By protecting wetlands, people save the plants that grow there.

◀ A park ranger in the Everglades National Park, Florida, helps students learn about the plants that grow there.

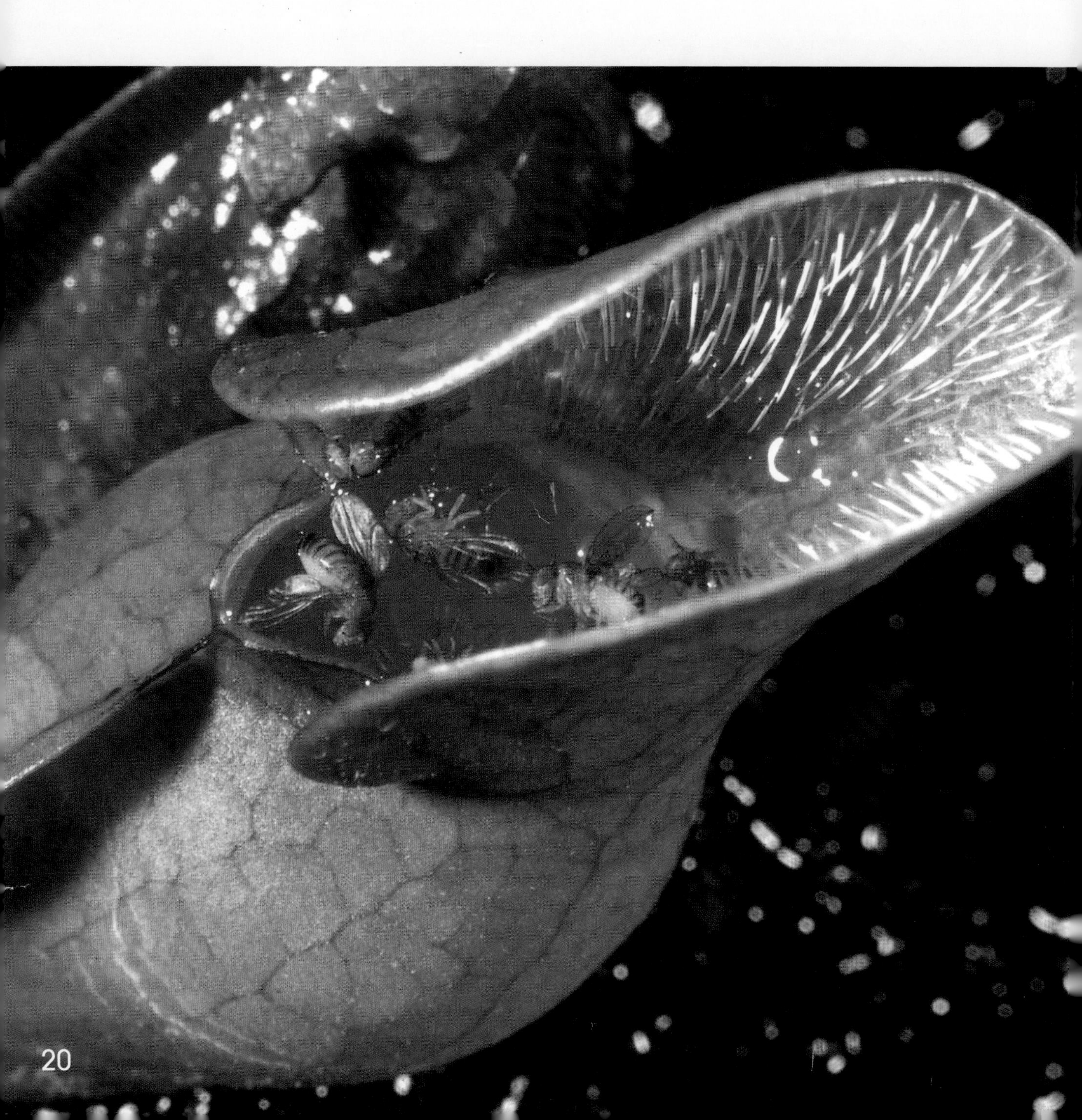

The Amazing Pitcher Plant

Insects need to be careful around the pitcher plant. This bog plant might eat them for lunch.

The pitcher plant has stiff hairs on the inside of its curved leaves. Insects easily travel down the stiff hairs. The plant is slippery where the hairs stop. Insects can't climb up the slippery wall. They fall into a pool of water in the leaves. The plant then slowly **digests** the insects. Insects give this plant the food it needs to live.

◄ Insects get trapped in water inside the pitcher plant. The plant uses insects for food.

Glossary

bog (BOG)—a type of wetland with wet, spongy land; bogs usually have a thick layer of peat.

digest (dye-JEST)—to break down food

marsh (MARSH)—a type of wetland; marshes usually have soft, moist soil covered in grasses.

mulch (MUHLCH)—a layer of sawdust, paper, or dead plants spread on soil to condition it

oxygen (OK-suh-juhn)—a colorless gas in the air that living things need to breathe

restore (ree-STOR)—to bring back to an original state

survive (sur-VIVE)—to continue to live

swamp (SWAHMP)—a type of wetland that usually has standing water and woody plant life

threatened (THRET-uhnd)—in danger of dying out

Read More

Kalman, Bobbie, and Amanda Bishop. *What Are Wetlands?* The Science of Living Things. New York: Crabtree, 2003.

Richardson, Adele D. *Wetlands.* The Bridgestone Science Library. Mankato, Minn.: Bridgestone Books, 2001.

Internet Sites

FactHound offers a safe, fun way to find Internet sites related to this book. All of the sites on FactHound have been researched by our staff.

Here's how:

1. Visit *www.facthound.com*
2. Type in this special code **0736843256** for age-appropriate sites. Or enter a search word related to this book for a more general search.
3. Click on the **Fetch It** button.

FactHound will fetch the best sites for you!

Index